AUTONOMOUS WHEELCHAIR

(Moving towards Intelligent Machines)

AUTHORS

Saima Rehman, Tabinda Ashraf,

Muhammad Umair, Osama Zubair,

Hammad Khan

ASSOCIATE PUBLISHER

AUTHORS UNITED

Page intentionally left blank

Dedicated To Our Beloved Parents and Teachers

Page intentionally left blank

COPYRIGHT STATEMENT

- Copyright in text of this book rests with the student authors. Copies (by any process) either in full, or of extracts, may be made **only** in accordance with instructions given by the author and lodged in the library of NUST College of E&ME. Details may be obtained by the librarian. This page must form part of any such copies made. Further copies (by any process) of copies made in accordance with such instructions may not be made without the permission (in writing) of the author.

- The ownership of any intellectual property rights which may be described in this book is vested in NUST College of E&ME, subject to any prior agreement to the contrary, and may not be made available for use by third parties without the written permission of the College of E&ME, which will prescribe the terms and conditions of any such agreement.

Page intentionally left blank

ACKNOWLEDGEMENTS

We start with the holy name of Allah with who gave us will, strength, and determination to complete this book.We are thankful to our supervisors Dr. Fahad Mumtaz Malik, Dr. Yasar Ayyaz and Mr. Hammad Khan for guiding us throughout the project and providing us with the material and information needed for the book. We express sincere gratitude to Mr. Ali Bin Wahid and Mr. Usama Siraj for helping us in understanding the hardware architecture and working of Pioneer 3-AT mobile robot. We gratefully acknowledge the assistance of Mr. Zaid Ahsan Shah for his valuable guidance, keen interest and encouragement at various stages of ourresearch. We thank all our seniors, friends, colleagues and mentors for their help and belief in us. Finally, we would like to express our sincere thanks to our parents. Without their prayers, we would not be able to complete this project.

Page intentionally left blank

ABSTRACT

Human-robot interaction is one of the most basic requirements for service robots. In order to provide the desired service, these robots are required to detect and track people in human cluttered environment. This book presents a novel approach for classifying a target person in a crowded environment. The system used the approach of Hough transform for human detection using camera mounted on wheelchair. Using data from Hough transform our system classifies the target person from other human beings in the environment. Our system tracks human being by gathering details of his position and velocity, and then converting this data into corresponding linear and angular velocity of wheelchair. System used in the project is an intelligent wheelchair with partial AI and partial human command based working architecture with LRF and stereo camera mounted on it. It works on a Linux OS and uses robot operating system (ROS) for the operation of the wheelchair. Our approach is feasible for mobile robots with an identical device arrangement.

Page intentionally left blank

TABLE OF CONTENTS

ABSTRACT _____ 9

INTRODUCTION _____ 15

   1.1 Structure of the book _____ 21

   1.2 Applications _____ 22

LITERATURE REVIEW _____ 25

   2.1 Related Research _____ 26

      2.1.1 Human Detection _____ 26

      2.1.2 Path Planning _____ 32

   2.2 Robotic Wheelchair Projects _____ 34

      2.2.1 Level and implementation of Autonomy _____ 35

      2.2.2 Human-Robot Interaction _____ 36

      2.2.3 Mechanics Most _____ 36

HARDWARE AND METHODOLOGY _____ 39

      3.1.1 Common Interfacing of Intelligent wheelchairs: _____ 41

         3.1.1.1 Joystick controlled operation _____ 42

         3.1.1.2 Available Alternatives of Joystick: _____ 42

         3.1.1.3 In Research _____ 42

      3.1.2 P3-AT Architecture _____ 43

         3.1.2.1 Why P3AT? _____ 44

      3.1.3 Wheelchair Design Amendments _____ 46

         3.1.3.1 Amendments In Existing Design Of Wheelchair _____ 47

         3.1.3.2 Addition of new Components _____ 47

      3.1.4 System Architecture _____ 48

         3.1.4.3 Laser Range Finder (LRF) _____ 52

         3.1.4.4 Stereo vision Camera _____ 53

         3.1.4.5 DC motors _____ 53

         3.1.4.6 Low level drive controller _____ 55

  3.1.4.7 User Interface _____ 56
  3.1.4.8 PC _____ 57

3.2 Software Design _____ 58

 3.2.1 WHY ROS? _____ 59
  3.2.1.1 Inter-platform operability _____ 59
  3.2.1.2 Modularity _____ 59
 3.2.2 Design Principles of ROS _____ 60
  3.2.2.1 Peer-To-Peer _____ 60
  3.2.2.2 Multi-Lingual communication _____ 60
  3.2.2.3 Open-Source _____ 60
 3.2.3 File System Level _____ 61
  3.2.3.1 Packages _____ 61
  3.2.3.2 Manifest _____ 61
  3.2.3.3 Stack _____ 61
  3.2.3.4 Stack Manifest _____ 61
  3.2.3.5 Message Types _____ 62
  3.2.3.6 Service Types _____ 62
 3.2.4 Computation Graph Level _____ 62
  3.2.4.1 Nodes _____ 62
  3.2.4.2 Master _____ 62
  3.2.4.3 Parameter Server _____ 63
  3.2.4.4 Messages _____ 63
  3.2.4.5 Topics _____ 63
  3.2.4.6 Services _____ 64
 3.2.5 ROS Workspace _____ 65
  3.2.5.1 Catkin Workspace _____ 65
  3.2.5.2 Rosbuild Workspace _____ 65
 3.2.6 Building the workspace _____ 66

3.3 Working _____ 66

 3.3.1 Image Acquisition and computer vision _____ 68
  3.3.2 RGB to HSV conversion _____ 70
  3.3.3 Image Smoothing _____ 72
  3.3.4 Thresholding _____ 74
  3.3.5 Object Detection Using Circular Hough Transform ____ 75

- 3.3.5.1 Amendments in the algorithm _____ 81
- 3.3.6 Obstacle Detection _____ 81
- 3.3.7 Obstacle Avoidance _____ 82
  - Basic Obstacle Avoidance _____ 82
  - Advanced Obstacle Avoidance _____ 83
- 3.3.8 Velocity Calculations _____ 83
  - 3.3.8.1 Angular motion _____ 83
  - 3.3.8.2 Linear motion _____ 84
  - 3.3.8.3 Obstacle avoidance _____ 85
- 3.3.9 Software Architecture _____ 86

3.4 Operation _____ 87
- 3.4.1 System Requirements _____ 87

EXPERIMENTS AND RESULTS _____ 91
- 4.1 Experimentation _____ 92
- 4.1.1 Type of motion of caretaker _____ 92
- 4.1.2 Indoor/outdoor _____ 93
- 4.1.3 Tele-operation using Android App _____ 93
- 4.1.4 Keyboard Teleop _____ 94

4.2 Limitations _____ 94

CONCLUSION _____ 95
- **5.1 Future Work** _____ 96

REFERENCES _____ 98

Page intentionally left blank

Chapter No. 1

INTRODUCTION

The development of service robots that are able to assist people in their daily tasks has become a popular research area in recent years. Possible application scenarios for such robots include support for the elderly and the general service robots for public areas or shopping malls. An essential prerequisite for service robots is the ability to recognize humans and interact with them in a non-technical, natural way.

The primary objective of this course is to promote the development of robust, repeatable and transferable software for robots that automatically detect, track and follow people around them. The work will be developed by the necessity of such functionality onboard on intelligent electric wheelchair robots to assist people with mobility impairments. Approach used in this book is effective for different robot platforms. The method has been implemented in the Robot Operating System (ROS) framework and will be publicly released as ROS package. This book based is project based with electric wheelchair as an example robotic platform.

The ability to autonomously detect, track and follow a person has been identified as an important functionality for many assistive and service-robot systems. In the last decade, significant advances have been made in the development of people detection and tracking algorithms, which are often carried out with the aim of improving the human-robot interaction and robot navigation in populated environments.

But most of the previous work is not easily transferable to new applications: algorithms are tested on a single robot in a single environment (if at all, sometimes only in the simulation under artificial conditions), in many cases, the code has not been made available for public, records collected in the validation sessions are not shared and quantitative comparisons to existing algorithms has not been performed.

In this project we focus on a human following as important, fundamental interaction capability for a service robot equipped with a color camera and a laser scanner.

The idea of intelligent wheelchair focuses on improving the lives of people with severe disabilities who are not in the situation of directing wheelchairs on their own. The aim of the project is to make wheelchair users able to autonomously navigate using advanced environment perception, motion planning and navigation techniques through everyday obstacle cluttered human environments. It offers multiple ways for control: totally un-autonomous (like ordinary wheelchairs); to move chair as a remote control moving platform, which can be commanded via android-teleop application and last is fully automated control, in which wheelchair follows the caretaker of the patient sitting on wheelchair, while taking care of the obstacles in the way.

Objectives of this book are:

- Developing in-house research platform for research and development in the field of Assistive Robotics.
    - Bringing in capabilities of Mobile Robots in Electric Wheelchairs.
- Reconfigurable Design
    - Developing a System considering the needs of User community.
    - Modular i.e. Flexibility in adding new sensors, interfaces or robotic functionalities.
- Human following capabilities

- Obstacle avoidance and path planning

This is an international project carried out in collaboration with Sakura Wheelchair project (Japan). Sakura wheelchair Project is the renowned wheelchair manufacturing company in Japan with the mission of manufacturing and provision of smart, active and disability based manual and electric wheelchairs to persons with disabilities to ensure their mobility for the creation of inclusive society for all. Sakura and Milestone launch project to provide use electric wheelchairs to persons with disabilities to ensure their mobility and independence. Sakura has provided SMME, NUST with one of their wheelchairs for research purposes to assist their cause. (Wheelchair costs 2.5 Million Rupees, if purchased)

A great deal of work has been done in this area and various methods have been applied to reach solution. Generally laser range finder and cameras have been the most common devices used for human detection. Some of the earlier methods used only laser data for human detection but it only provides 2D information. Modern approaches use both laser and camera fusion techniques. SongminJia et al. [1] present a human detection method based on extraction of human model from disparity image by stereo camera. Further improvement in the robustness and real-time performance of their algorithm has been presented in [2]. Mehrez Kristou et al. [3] present an approach that uses omnidirectional camera and laser range finder to detect human being. Their approach uses human's cloth pattern to identify the target person and the process consists of two stages: registration stage and identification and localization stage. They further improved their approach by merging the registration stage in human tracking phase [4]. Nicola Bellotto and Huo sheng Hu used laser range finder to detect human legs and PTZ camera to find faces [5]. They used Unscented Kalman filter approach to perform real-time tracking. Orasa Patsaduet al. [6] presents another approach for recognition of human gesture using Kinect camera.

Generally speaking, wheelchair is capable of sensing its caretaker, location, interpreting the sensed information to obtain the knowledge of its location and direction of motion, planning a real-time trajectory to reach the object. In this process, the issue of obstacle avoidance is a fundamental topic to be challenged. It is one of the most fundamental problems that have to be solved before the mobile robot can navigate and explore autonomously in complex indoor environment. The aim of path planning is to generate a feasible path which can guarantee mobile robot's moving from start to the target safely and optimally.

In wheelchair project, we have utilized data from laser range finder to identify obstacle and move along a path to avoid obstacle.

1.1 Structure of the book

This book is arranged and compiled with the aim to facilitate the reader in understanding every basic detail of the project. Chapter 1 draws the attention of the reader towards the problem which we have recognized in this project and the proposed solution. The chapter is an introduction to the book and applications of the project. Moving forward, chapter 2 provides the background knowledge necessary for reader, which makes a solid base because of which this project can be understood easily. After getting the knowledge of where the world is going, new technologies and new trends in research; in chapter 3 we are now in position to work on the wheelchair. This chapter starts with detailed description of hardware architecture of the Wheelchair, and then moves forward to software details. This chapter also comprises of the working of wheelchair, so that even a totally new person who has no idea of the project, can also operate it. Multiple experiments that we did for checking the performance of wheelchair are listed in chapter 4 along with results. Finally comes the conclusion.

1.2 Applications

Though all the algorithms and techniques used in the project refer to the problem of accurate tracking and following of people by social and assistive robots, we expect our work to have significant applications outside of this field, including for security (e.g. tracking intruders), entertainment (e.g. developing interacting exhibits), marketing (e.g. location-aware personalized advertisement), rehabilitation (e.g. assessment of patients' locomotion patterns following an injury), and beyond. This project is equally applicable on any mobile robot with a laser scanner near its base and a camera mounted on the top.

Page intentionally left blank

Chapter No. 2

LITERATURE REVIEW

2.1 Related Research

Interaction between humans and robots is a fundamental need for assistive and service robots. Their ability to detect and track people is a basic requirement for interaction with human beings. A long list of researches have been done in this field utilizing a variety of techniques and algorithms. We will discuss those in two categories i.e.

- Human detection and tracking
- Path planning

2.1.1 Human Detection

A large amount of research has been carried out on human tracking robots. Generally, a laser rangefinder and camera have been the most commonly-used sensors. Jia et al. [7] presented a human detection and tracking approach using stereo vision and extended Kalman filter-based methods. Kristou et al. [8] presented their approach for identifying and following the targeted person using a laser rangefinder and an omnidirectional camera. Their method fused the data of both these sensors by identifying the target using a panoramic image, and then tracking that targeted person using the laser rangefinder. Bellotto et al. [9] also presented their multi-sensor fusion system to track people using a laser range sensor and PTZ camera. They used laser scans to detect human legs and video images to detect the human face. They further used an unscented Kalman filter to generate multiple human tracks. People recognition based on their gestures is presented by Patsaduet al. [10]

Recently, Petrović et al. [11] presented a real-time vision-based tracking method using a modified Kalman filter. They used a stereo vision-based detection method to get the features from 2D stereo images, and then reconstructed them into 3D object features to detect human beings. This technique is suitable when only one person is present in the environment – the target person – and it is not able to handle more than one person in the environment.

Schulz et al. [12] proposed to estimate the number of people in the current scan based on the number of moving local minima in the scan. Unfortunately, this requires people move continuously to be tracked, and is susceptible to poor results in cluttered environments (where the number of local minima is misleading). They also introduced a Sample Based Joint Probabilistic Data Association Filter (SJPDAF) over the observed local minima to improve tracking reliability.

Topp et al. [13] extended [12], by picking out shapes of legs and person-wide blobs in laser scans using hand-coded heuristics, to allow detection and tracking of both stationary and moving people. The approach was also combined with a person following navigation algorithm, combining both the tracked person's position as well as the location of nearby obstacles to determine suitable control. Gockley et al. [14] used a similar approach, with a few modifications, including using a Brownian motion model for the tracking component. This approach was further extended by Hemachandra [15], which improved the person-following component by proposing a navigation approach that accounts for personal space, while avoiding obstacles. Unfortunately these approaches cite tracking difficulties in cluttered conditions, since they relied primarily on detecting clusters of a heuristically determined size in the laser scan.

More recently, Arras et al.[16] reduced this limitation by proposing a method that detects legs by first clustering scan points and then using supervised learning to learn shapes of leg clusters. Detected legs are tracked over time using constant-velocity Kalman filters and a multiple-hypothesis tracking (MHT) data association technique. This approach benefits from its ability to maintain (but not initiate) tracks of stationary people, and does not require any priori occupancy grid map of the environment. Initial results for this method appear promising, but demonstrations on walking speed robots in cluttered and crowded areas have yet to be performed. Thus, many questions remain about the robustness and generalizability of the approach.

Kobilarov et al. [17] developed a person-following Segway robot using an Omni-directional camera and a laser scanner. Nonetheless, variable environment lighting, backgrounds and appearances of people are factors which can be difficult to control for and can mislead vision-reliant tracking systems.

Fast and accurate detection of circles is widely applied in the fields of image processing and computer vision. The

Hough transform (HT), as a basic method to detect circles, has its advantages in insensitivity to noise in images and easiness in parallel computing, which has attracted extensive research among researchers. HT has the advantages of small storage and high speed if the parameter space is limited to two dimensions. Based on this we decided to detect the caretaker based on a circular tag present of his/her shirt rather than detecting the person himself.

Sambarta Dasgupta [18] presented an algorithm for the automatic detection of circular shapes from complicated and noisy images without using the conventional Hough transform methods. The proposed algorithm is based on a recently developed swarm intelligence technique, known as the bacterial foraging optimization (BFO). A new objective function has been derived to measure the resemblance of a candidate circle with an actual circle on the edge map of a given image based on the difference of their center locations and radii lengths. Guided by the values of this objective function (smaller means better), a set of encoded candidate circles are evolved using the BFO algorithm so that they can fit to the actual circles on the edge map of the image. The proposed method is able to detect single or multiple circles from a digital image through one shot of optimization.

T. D`Orazio [19] introduced a circle detection algorithm based on the CHT that has been formulated as convolutions applied on the edge magnitude image. The convolution kernels have been properly defined in order to detect the most completed circle in the image, being independent on the edge magnitude, and considering also different shapes of the ball according to different light conditions.

In 2005, Victor Ayala-Ramirez [20] presented a circle detection method based on genetic algorithms. It reduces the search space by avoiding trying unfeasible individuals, this result in a fast circle detector. The approach detects circles with sub-pixellic accuracy on synthetic images, but it can also detect circles on natural images. It also handles detection of occluded circles and several circles in single image.

T. J. Atherton [21] showed that a specific combination of modification to the CHT is formally equivalent to applying a scale invariant kernel operator.

2.1.2 Path Planning

In recent years, designs of autonomous car-like robots have received increased attention due to their potential applicability and usefulness in the automotive and robotics industry. The autonomy of robots depends on the capability of the robots to explore unknown environments. Different areas of research, such as parking, navigation, trajectory tracking, and wall/lane-following, are all actively being investigated. Nowadays, mobile robots are a promising technology that will improve the quality of life by providing collision avoidance, reducing traffic gridlock, and allowing the replacement of dangerous tasks currently performed by human drivers. The path planning for the objective of obstacle avoidance which is one of the key issues in the mobile robot navigation is the major topic to be investigated in this study. In general, the path planning can be divided into two categories: global path planning and local path planning. In the global path planning, the prior knowledge of the robot workspace should be available. In local path planning methods use ultrasonic sensors, laser range finders, and on-board vision systems to perceive the environment to perform on-line path scheduling. In this study, the workspace for the navigation of the mobile robot is assumed to be unknown and it has stationary obstacles only.

Most previous work focuses on only one or two of the three identified sub-problems: detection, tracking and following. Few papers present integrated systems tackling all three components. We focus in particular on work that uses depth sensor, e.g. laser range finder, and visual camera, as these are the sensosr available on our Smart Wheeler robot, as well as on numerous other assistive robots.

Monte Merlo et al. [22] were possibly the first to present a method for automatically locating and tracking people from laser data. One drawback of this work is the necessity for an a priori occupancy grid map of the operational environment, which is used for background subtraction to detect the person. The tracking is achieved via a conditional particle filter.

Wang and Liu [23] resolved the local navigation problem by recording the information of obstacle and trajectory into a "memory grid" of the environment. The grid-based map represents the robot workspace by a two dimensional array of elements as cells. However, it will spent sizable memory to store the robot workspace.

2.2 Robotic Wheelchair Projects

Many groups have proposed or developed robotic wheelchairs. A multitude of different techniques and ideas have been implemented or discussed. Each group has proceeded according to their focus or interest and these factors have varied widely. This section summarizes the main areas of variance between these projects. Such a survey of existing robotic wheelchairs serves to isolate and demonstrate fundamental design issues.

2.2.1 Level and implementation of Autonomy

The level of autonomy provided by a robotic wheelchair is an important evaluation criterion. At one extreme are manual designs that do not provide any control assistance to the operator. It may be argued that these are not robotic wheelchairs at all. However, they can still involve sophisticated processing, require specialized hardware and present complex control problems. [24]

Specialized techniques are used to support higher levels of autonomy. Some robotic wheelchairs make decisions based on internal occupancy grids [25]. These data structures organize sensor information into a coarse map, which is then used to inform speed and direction decisions. Other systems utilize fuzzy logic and a specialized rule base [26].

The Maid wheelchair [27] is highly autonomous. Its specialty is movement through crowded, changing environments. In one test the device crossed the floor of a busy railway station without colliding with anyone or anything.

2.2.2 Human-Robot Interaction

Robotic wheelchairs, by their nature, demand specialized user interfaces. Whilst many projects do not address this issue directly, invariably using joysticks [28], others do. A design [29] uses natural language commands, such as "move forward" or "move left", though a headset microphone.

The use of EEG measurements and signal processing to form a direct brain-computer interface for those suffering severe motor-related impairments have been suggested as a means of controlling a wheelchair [30]. However, this research is still far from practical application.

2.2.3 Mechanics Most

Robotic wheelchairs are implemented by modifying existing Powered Wheelchair systems. Examples include the Bremen Autonomous Wheelchair [12] and the Maid (Mobility Aid for Elderly and Disabled people) robotic vehicle [13]. These approaches arrange sensors and computing hardware around an existing infrastructure. They are able to take advantage of pre-built control and motor systems. One such project, the Tin Man wheelchair [14], uses servomotors to control the host chair through an unmodified joystick.

Page intentionally left blank

Chapter No. 3

HARDWARE AND METHODOLOGY

3.1 Hardware Architecture

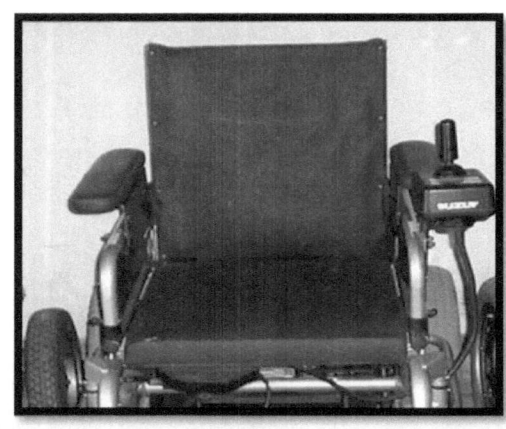

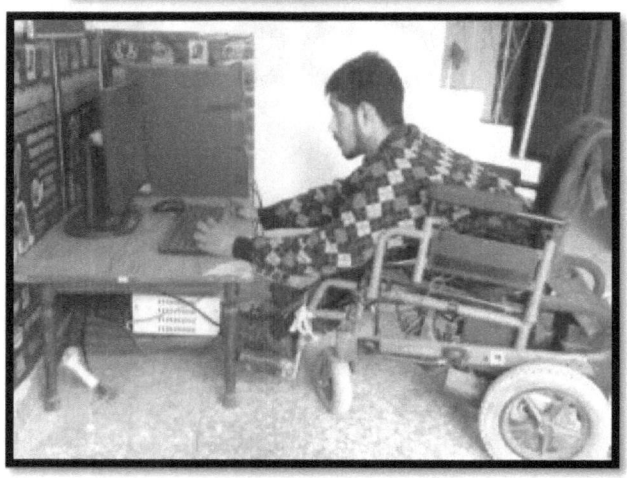

Figure 1: Sakura Wheelchair Project

Sakura Wheelchair Project is the renowned wheelchair manufacturing company in Japan with the mission of manufacturing and provision of smart, active and disability based manual and electric wheelchairs to persons with disabilities to ensure their mobility for the creation of inclusive society for all. Sakura and Milestone launch project to provide use electric wheelchairs to persons with disabilities to ensure their mobility and independence.

Sakura has given a Wheelchair to SMME, NUST for research purpose, to ensure better life for disables. (The actual cost of this wheelchair is 2.5 million rupees including delivery charges.) NUST added Laser range finder, camera, charging circuitry and PC.

3.1.1 Common Interfacing of Intelligent wheelchairs:

The wheelchairs in research have been made semi-autonomous and possess various features that are being implemented. The wheelchair we acquired had similar features. It was then made and hardware designed according to our needs of making it an autonomous human following wheel chair.

3.1.1.1 Joystick controlled operation

The commonly seen user interfacing of the wheelchairs up till now include the joystick interface. Joystick allows occupants to specify velocity and rate of rotation. The joystick x-axis is interpreted as the desired rate of rotation and the y-axis as the desired velocity.

3.1.1.2 Available Alternatives of Joystick:

The available alternatives for the navigation of wheelchair include, Touch, flex, Sip-n-Puff, Chin-controlled, head-controlled, speech controlled, tongue controlled etc.

3.1.1.3 In Research

In research we see many engineers working on the Eye-movement, EEG, EMG, EOG, and Visual Gesture based movement and control of t intelligent wheelchairs.

3.1.2 P3-AT Architecture

Pioneer is a family of mobile robots, both two-wheel and four-wheel drive, including the Pioneer 1 and Pioneer AT, Pioneer 3-AT and many others. These small, research and development platforms share common architecture and foundation software with all other MOBILEROBOTS platforms. MOBILEROBOTS platforms set the standards for intelligent mobile robots by containing all of the basic components for sensing and navigation in a real-world environment.

The platform comes to complete with a sturdy aluminum body, balanced drive system four-wheel skid-steer, reversible DC motors, motor-control and drive electronics, high-resolution motion encoders, and battery power, all managed by an onboard microcontroller and mobile-robot server software.

Besides the open-systems robot-control software onboard the robot microcontroller has a host of advanced robot-control client software applications and applications-development environments. Software development includes our own foundation Advanced Robotics Interface for Applications (ARIA) and AR Networking, with fully documented C++, Java and Python libraries and source code.

3.1.2.1 Why P3AT?

i. Easy to Use - Comes fully assembled and integrated with its accessory packages.
ii. Reliable - Construction is durable and rugged. Easily handles the small gaps, minor bumping, jarring, or other obstacles that hinder other robotic platforms. Some Pioneer robots have been in service for over 15 years.
iii. Pioneer Software Development Kit - All Adept Mobile Robots platforms include Pioneer SDK, a complete set of robotics applications and libraries that accelerate the development of robotics projects. Pioneer SDK is backed by our product support team.
iv. Customizable - Easily accessorize by choosing from dozens of supported and tested accessories that integrate with your robotic platform. Additional

help is available for future upgrades or added accessories.

v. Reference Platform - Pioneer robots are a standard in intelligent mobile platforms. Search your preferred robotics journal or conference listings to find many examples of Pioneer platforms in research applications.

vi. Technical Support - Pioneer software and hardware comes fully documented with additional help available through our product support team

3.1.3 Wheelchair Design Amendments

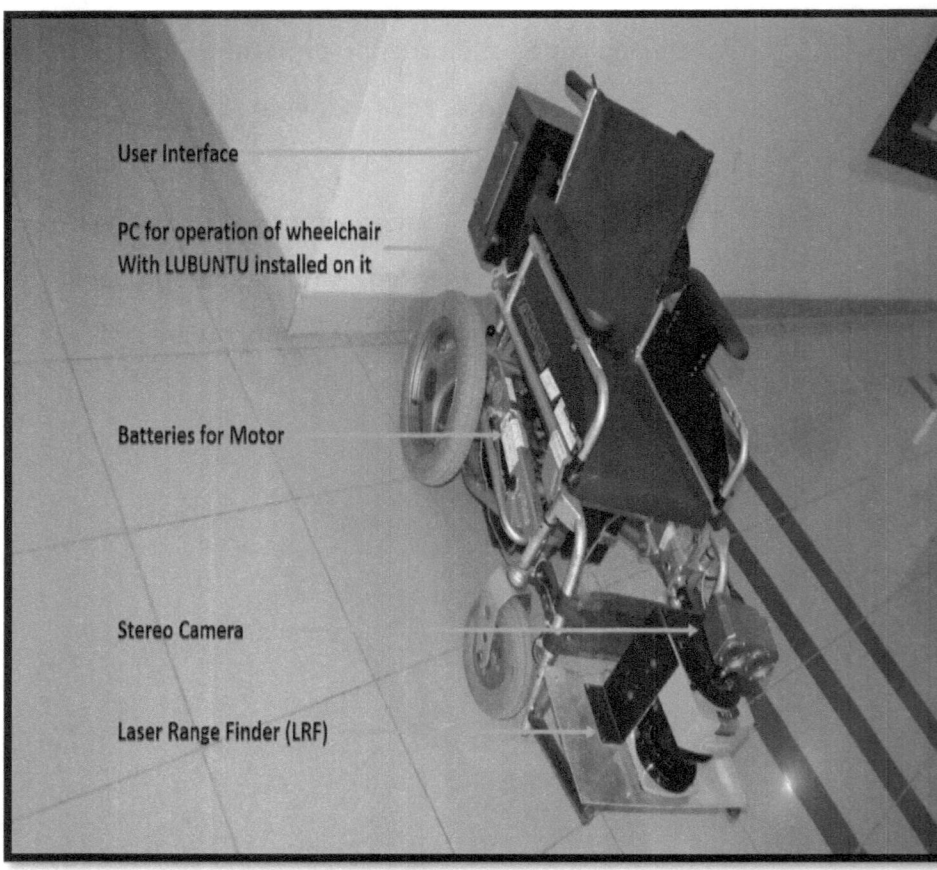

Figure 2 Wheelchair Architecture with all parts labeled

We amended design of wheelchair and added new components from Pioneer3AT and other sources. Our approach of making it fully autonomous and DIP based human following needed a few hardware amendments as follows:

3.1.3.1 Amendments In Existing Design Of Wheelchair

- Joystick knob is unplugged from the potentiometers, which were generating voltage commands.
- The voltage signals corresponding to the movement of wheelchair wheels were then decoded.
- The wheelchair was programmed to be controlled accordingly.

3.1.3.2 Addition of new Components

- Laser Range Finder (Sick LMS200)
- Stereo Vision Camera
- Wheelchair Computer

Hardware Architecture thus includes:

Instrumentation	Cobra Vesalogic SBC, SICK LMS 200 LRF, Mobile Ranger Stereo Vision Camera
Devices/OS	Android Tablet, Linux Computer
User Interfaces	USB Joystick, Mouse, Keyboard, Emotive Headset

So the design has p3AT architecture mounted on it. It further has multiple sensors whose data fusion is taken into use for achieving the purpose of human detection and obstacle avoidance.

3.1.4 System Architecture

Figure 3 Rear view of wheelchair

48

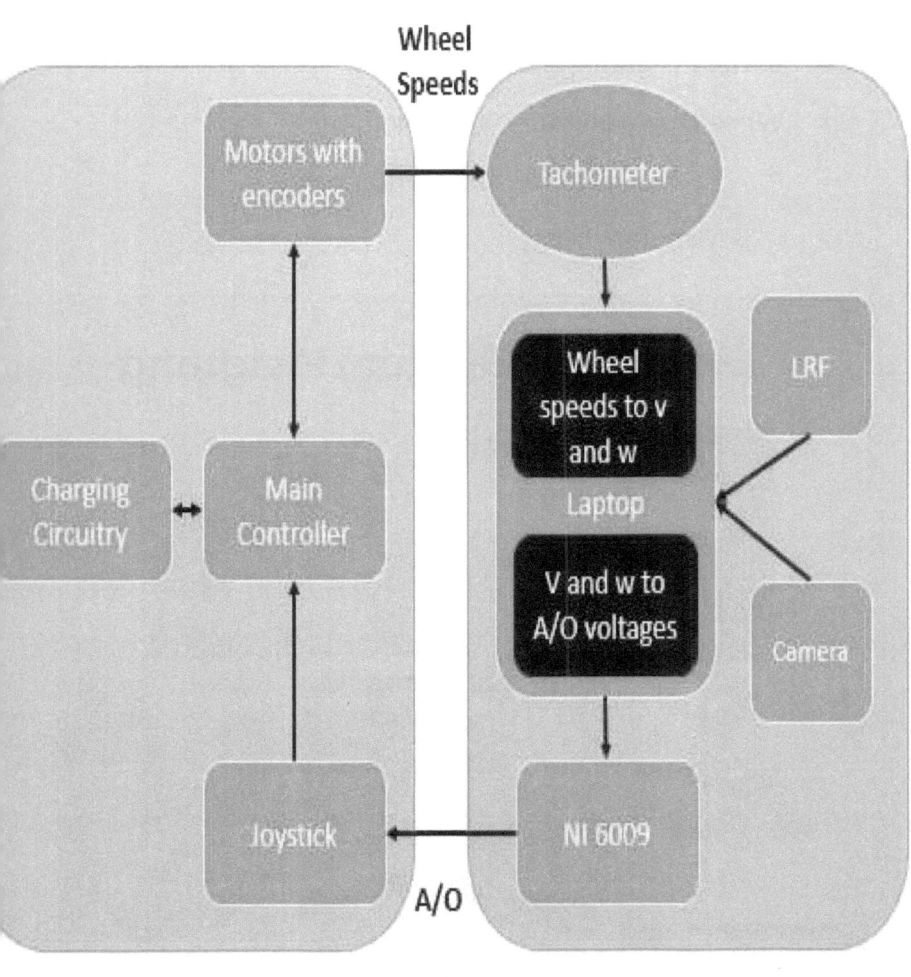

Figure 4 System Architecture

3.1.4.1 Electronic Circuit Designing

- Battery charging circuitry
- Wheelchair communication circuitry
- RS-232 interfaces for sensor interfacing
- PCB Manufacturing

Figure 5 Electronic circuit designing

3.1.4.2 CAD Modeling:

- Wheelchair model
- Mounting Models

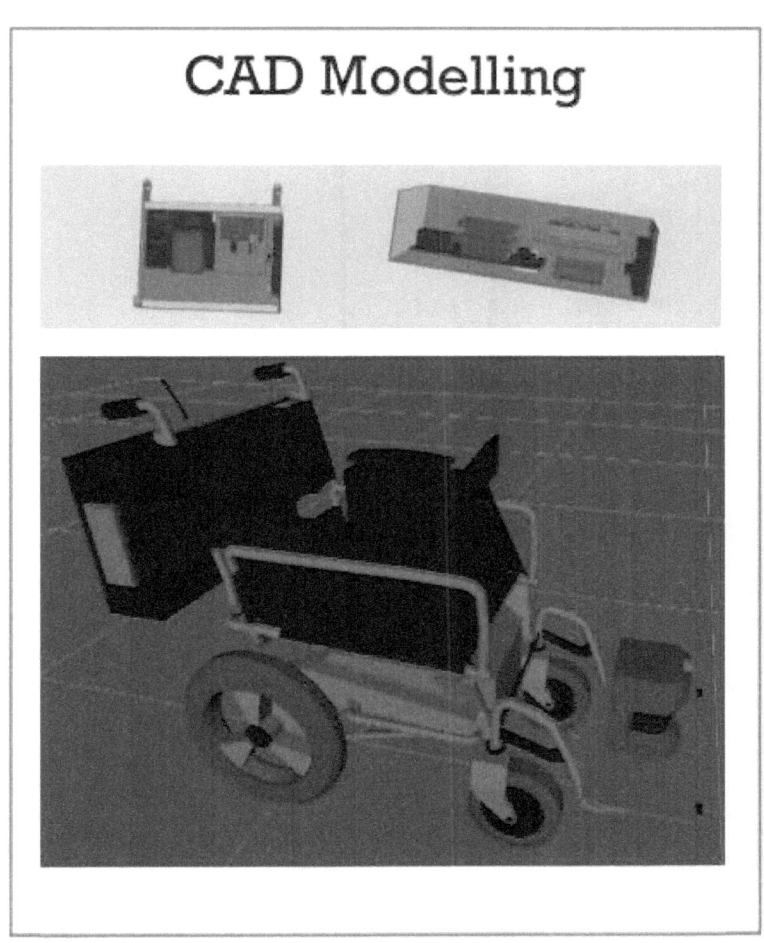

Figure 6 CAD Modelling

3.1.4.3 Laser Range Finder (LRF)

The LMS200 SICK LRF used is a Laser Measurement System which is a non-contact measurement systems used in various applications. It scans the surroundings two-dimensionally with a radial field of vision using infra-red laser beams (laser radar).

These are used for:

- Area monitoring
- Determining positions
- Object measurement and detection

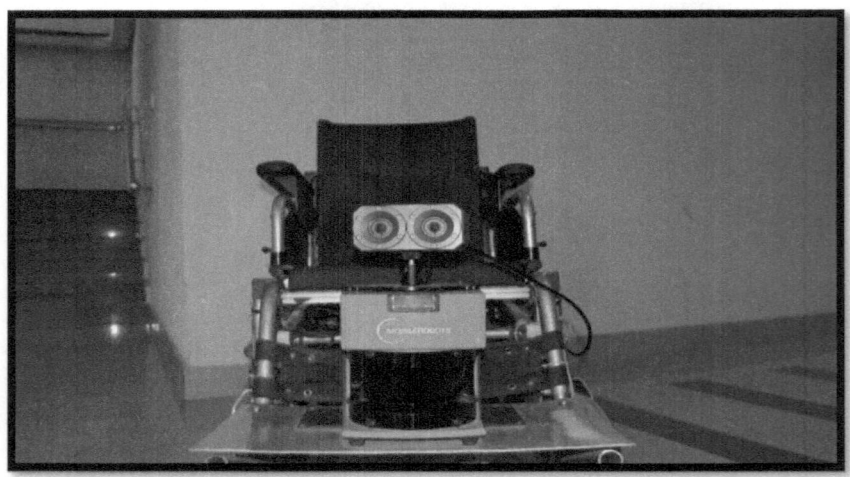

Figure 7 Camera and laser range finder

3.1.4.4   Stereo vision Camera

Stereo vision works in a similar way to 3D sensing in human vision. It begins with identifying image pixels that correspond to the same point in a physical scene observed by multiple cameras. The 3D position of a point can then be established by triangulation using a ray from each camera. The more corresponding pixels identified, the more 3D points that can be determined with a single set of images. Correlation stereo methods attempt to obtain correspondences for every pixel in the stereo image, resulting in tens of thousands of 3D values generated with every stereo image.

3.1.4.5 DC motors

2 DC motors with encoders are used in the hardware for the two rear wheels. DC motors instead of servo are used because of multiple reasons:

DC Motor:

- High output power relative to motor size and weight.
- Encoder determines accuracy and resolution.

- High efficiency. It can approach 90% at light loads.
- High torque to inertia ratio. It can rapidly accelerate loads.
- Has "reserve" power. 2-3 times continuous power for short periods.
- Has "reserve" torque. 5-10 times rated torque for short periods.
- Motor stays cool. Current draw proportional to load.
- Usable high speed torque. Maintains rated torque to 90% of NL RPM
- Resonance and vibration free operation.

Servo Motor:

- Low efficiency. Motor draws substantial power regardless of load.
- Torque drops rapidly with speed (torque is the inverse of speed).
- Low accuracy. 1:200 at full load, 1:2000 at light loads.

- Prone to resonances. Requires micro stepping to move smoothly.
- No feedback to indicate missed steps.
- Low torque to inertia ratio. Cannot accelerate loads very rapidly.
- Motor gets very hot in high performance configurations.
- Motor will not "pick up" after momentary overload.
- Motor is audibly very noisy at moderate to high speeds.
- Low output power for size and weight.

3.1.4.6 Low level drive controller

Here are few specifications of the used controller:

- System Serial
- 32 digital inputs
- 8 digital outputs
- 7 analog inputs
- 3 serial expansion ports

3.1.4.7 User Interface

Figure 8 User Interface

The User interface is a simple and efficient interface with switch buttons for the operations. The buttons include:

AUX1: Connection of wheelchair PC with its rechargeable batteries.

AUX2: Connection of wheelchair with the Wi-Fi device.

Motors: Turning the switch on, turns on the DC motors.

ON/OFF: It is the Main power switch of the PC.

Other connections include ETHERNET interface, Keyboard, Mouse and Charging slot.

3.1.4.8 PC

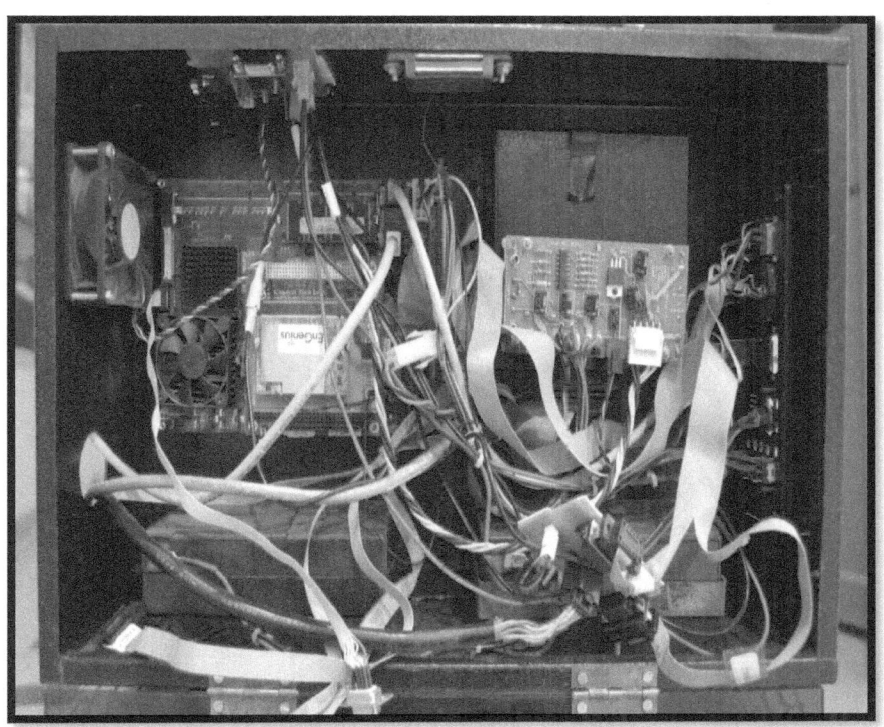

Figure 9 PC with LUBUNTU installed on it

The computer is mounted at the rear of wheelchair in a computer Rack. The operating system is currently a Linux computer i.e. Lubuntu (Light Ubuntu). Lubuntu is a fast and lightweight operating system. The core of the system is based on Linux and Ubuntu. Lubuntu uses the minimal desktop LXDE, and a selection of light applications. The focus is on speed and energy-efficiency. Because of this, Lubuntu has very low hardware requirements.

3.2 Software Design

The core software used for the programming is ROS (Robots Operating System).It is a flexible framework for ROS is an open-source, meta-operating system for your robot. It provides the services you would expect from an operating system, including hardware abstraction, low-level device control, implementation of commonly-used functionality, message-passing between processes, and package management. It also provides tools and libraries for obtaining, building, writing, and running code across multiple computers.

3.2.1 WHY ROS?

3.2.1.1 Inter-platform operability

ROS message-passing means that you can work between very different components and subsystems that are probably running with different languages (maybe something like low-level hardware control with C for speed, and high-level state machines with Java or Python for ease of coding). This also gets around the problem of the mess of APIs you would have had to deal with before.

3.2.1.2 Modularity

Since things are connected by a distributed message system, if one component crashes, your whole system doesn't crash. Granted, there are plenty of ways to make your system more robust so that this doesn't happen in the first place, but ROS makes it easier for your robot to continue doing its thing even if two sensors and an arm motorare no more functional.

3.2.1.3 Concurrent Resource Handling

Without ROS, reading/writing to resources quickly becomes a mess with large multi-threaded systems (i.e. virtually any robotics application). Again, there are ways to deal with this, but ROS simplifies the whole process by ensuring that your threads aren't actually trying to read and write to shared resources, but are rather just publishing and subscribing to messages.

3.2.2 Design Principles of ROS

3.2.2.1 Peer-To-Peer

ROS works in a manner that there are several independent processes and hosts working together, integrated and sharing tasks with one another.

3.2.2.2 Multi-Lingual communication

The communication is based on XML-RPC supports C++, Python, Octave, LISP, Java defined data types, proprietary message definition files Tool-Based Linux philosophy with small building blocks.

3.2.2.3 Open-Source

The source code are made available with license in which copyrights are reserved. It has been developed in a collaborative public manner.

3.2.3 File System Level

3.2.3.1 Packages

Packages are the main unit for organizing software in ROS. A package may contain ROS runtime processes (*nodes*), a ROS-dependent library, datasets, configuration files, or anything else that is usefully organized together.

3.2.3.2 Manifest

Manifests (manifest.xml) provide metadata about a package, including its license information and dependencies, as well as language-specific information such as compiler flags.

3.2.3.3 Stack

Stacks are collections of packages that provide aggregate functionality, such as a "navigation stack." Stacks are also how ROS software is released and have associated version numbers.

3.2.3.4 Stack Manifest

Stack manifests (stack.xml) provide data about a stack, including its license information and its dependencies on other stacks.

3.2.3.5 Message Types

It defines the data structure for messages.

3.2.3.6 Service Types

It defines the request and response data structure for services.

3.2.4 Computation Graph Level

3.2.4.1 Nodes

Nodes are processes that perform computation. ROS is designed to be modular at a fine-grained scale; a robot control system will usually comprise many nodes. For example, one node controls a laser range-finder, one node controls the wheel motors, one node performs localization, one node performs path planning, and one Node provides a graphical view of the system, and so on. A ROS node is written with the use of a ROS client library, such as roscpp or rospy.

3.2.4.2 Master

The ROS Master provides name registration and lookup to the rest of the Computation Graph. Without the Master, nodes would not be able to find each other, exchange messages, or invoke services.

3.2.4.3 Parameter Server

The Parameter Server allows data to be stored by key in a central location. It is currently part of the Master.

3.2.4.4 Messages

Nodes communicate with each other by passing messages. A message is simply a data structure, comprising typed fields. Standard primitive types (integer, floating point, Boolean, etc.) are supported, as are arrays of primitive types. Messages can include arbitrarily nested structures and arrays (much like C structs).

3.2.4.5 Topics

Messages are routed via a transport system with publish / subscribe semantics. A node sends out a message by *publishing* it to a given topic. The topic is a name that is used to identify the content of the message. A node that is interested in a certain kind of data will *subscribe* to the appropriate topic. There may be multiple concurrent publishers and subscribers for a single topic, and a single node may publish and/or subscribe to multiple topics. In general, publishers and subscribers are not aware of each other's existence. The idea is to decouple the production of information from its consumption. Logically, one can think of a topic as a strongly typed message bus. Each bus has a name, and anyone can connect to the bus to send or receive messages as long as they are the right type.

3.2.4.6 Services

The publish/subscribe model is a very flexible communication paradigm, but its many-to-many, one-way transport is not appropriate for request/reply interactions, which are often required in a distributed system. Request / reply is done via services, which are defined by a pair of message structures: one for the request and one for the reply. A providing node offers a service under a name and a client uses the service by sending the request message and awaiting the reply. ROS client libraries generally present this interaction to the programmer as if it were a remote procedure call.

3.2.5 ROS Workspace

It is a catkin workspace in which one or more packages can be built. There are two types of workspaces in ROS:

3.2.5.1 Catkin Workspace

A catkin workspace is a folder where you modify, build, and install catkin packages. Catkin packages can be built as a standalone project and also provides the concept of workspaces, where multiple interdependent packages can be build, together all at once.

3.2.5.2 Rosbuild Workspace

Rosbuild is supported for some more time, given the huge amount of packages out there built with rosbuild as compared to the catkin workspace. Rosbuild packages are also called dry packages, because the introduction of catkin to existing ROS packages starts with low-level (core) packages and slowly moves up to higher-level packages, like a rising tide.

3.2.6 Building the workspace

Rosmake is the command used to build the ROS packages. The command displays all the paths used for each spacing. The folders get created in the catkin workspace.

3.3 Working

Here is the workflow for this project:

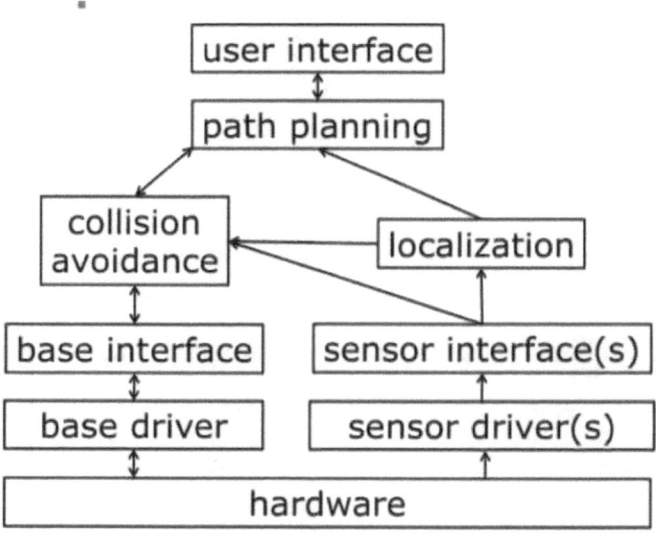

Figure 10 Integration of different capabilities

This is the layout of how we have proceeded with the project.

- Took real time image data from camera
- Processed the obtained image for recognizing caretaker
- Took real time data from Laser Range Finder
- Processed the received data to retrieve obstacle information
- Fused both of the extracted results to predict details of the surrounding
- In case if there are any obstacles in the way to caretaker, new path is planned that wheelchair has to follow, while keeping the track of its caretaker.
- Finally velocity commands are issued to wheelchair which moves it in the desired direction with required speed.

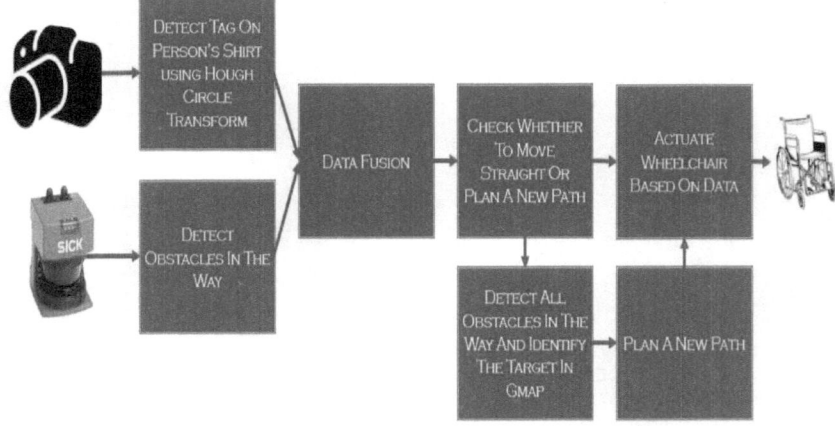

Figure 11 Overview of working of wheelchair

Now all of these steps are explained in very detail in upcoming sections.

3.3.1 Image Acquisition and computer vision

The first stage of any vision system is the image acquisition stage. The image that is acquired is completely unprocessed and is the result of whatever hardware was used to generate it. After the image has been obtained, various methods of processing can be applied to the image to perform the many different vision tasks required. Here are few of the processing techniques that we have applied:

- RGB to HSV conversion
- Image Smoothing
- Thresholding
- Hough Circle Transformation

In order to detect a human, we had various options available:

- Face detection
- Human detection
- Upper body detection
- And many more

But we didn't choose any of these because these detection/recognition techniques are computationally extensive and the processing time becomes an issue. That time taken by this processing is so much that the detection technique can't be applied to frames of video stream in real time. So in order to save computational time, we have decided to detect a particular tag present on the shirt of caretaker, for simplicity the tag is assumed to be a smooth circle of red color.

3.3.2 RGB to HSV conversion

Images can be processed in either RGB (Red, Blue, and Green) format or the HSV (Hue, Saturation, and Value) model. HSV is used when the color description has to play an important role. It describes colors in a way similar to how a human eye tends to recognize colors. It uses the familiar comparisons for description like color, vibrancy and brightness.

- Hue represents the color type its value is normalized to a range from 0 to 255, with 0 being red.
- Saturation represents the vibrancy of the color. Its value ranges from 0 to 255. The lower the saturation value, the more gray is present in the color, causing it to appear faded.
- Value represents the brightness of the color. It ranges from 0 to 255, with 0 being completely dark and 255 being fully bright.

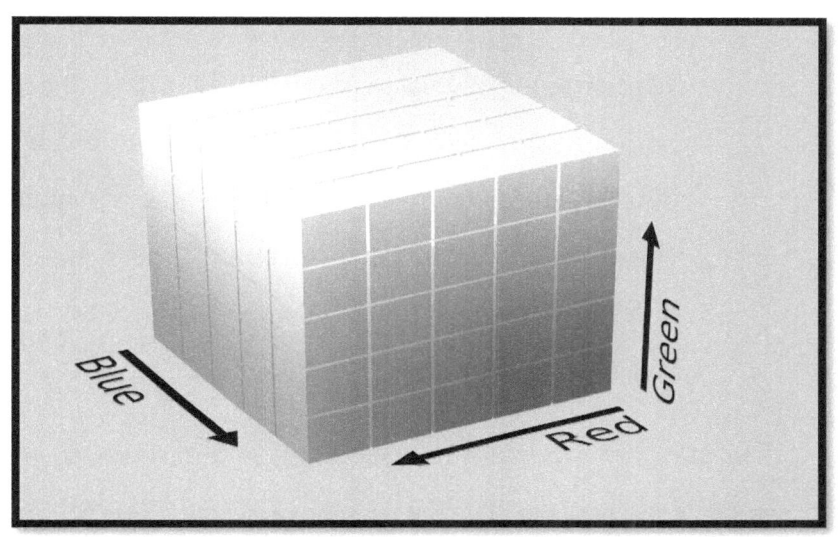

Figure 12 RGB color model

Figure 13 HSV color model

- White has an HSV value of 0-255, 0-255, 255. Black has an HSV value of 0-255, 0-255, 0. The dominant description for black and white is the term, value. The hue and saturation level do not make a difference when value is at max or min intensity level.

The color camera uses the RGB model to determine color. Once the camera has read these values, they are converted to HSV values. The HSV values are then used in the code to determine specific color. The pixels are individually checked to determine if they match a predetermined color threshold.

3.3.3 Image Smoothing

A simple and frequently used OpenCv operation is image smoothing. It is also called blurring the image. We need to blur or smoothen the image to reduce the noise in the image.

For smoothing an image, we apply a filter; mostly the filters used are the linear filters, in which the weighted sum of input pixels determines the output pixels.

$$g(i,j) = \sum_{k,l} f(i+k, j+l) * h(k,l)$$

Where:

f = Original Image

h = Kernel

g = filtered image

The second term in the summation is the kernel i.e. the coefficients of the filter. This is helpful in visualizing the filter as a sliding window of coefficients over the image.

The most commonly used filter that we implemented is the Gaussian Filter. The filter works by convolving each point in the input array with the Gaussian Kernel followed by the addition of all to give the Output array.

3.3.4 Thresholding

Now since the noise content has been suppressed in the image it can be processed further. The first processing that we do is thresholding. As due to reflection and brightness color value is obtained is not always same. So we define a range of HSV values for our tag, and threshold the image:

$$f(x,y) = \begin{cases} 1 & if\ HSV(x,y) \in HSV\_tag \\ 0 & otherwise \end{cases}$$

Here a 1 corresponds to white color, while a 0 represents black color. HSV(x,y) is the HSV value of the pixel being processed while HSV_tag is the defined range of HSV values for tag.

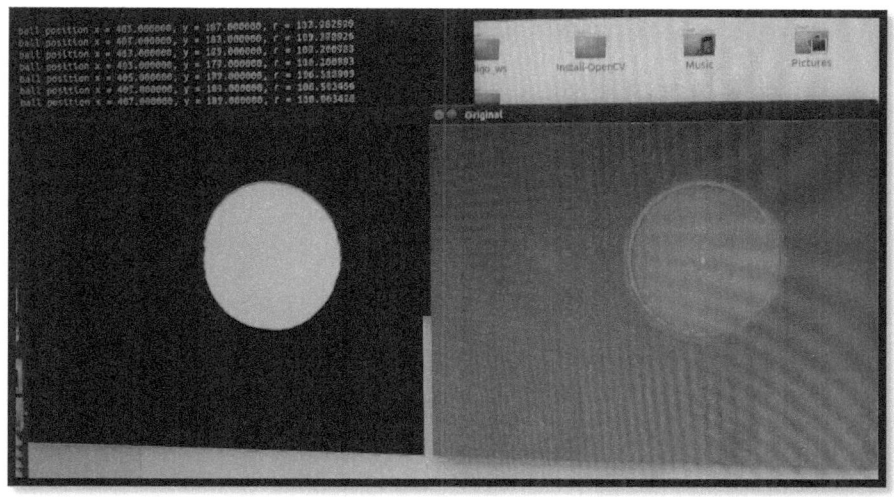

Figure 14 circle detection using Hough transform technique

3.3.5 Object Detection Using Circular Hough Transform

Among the most challenging techniques of image processing and computer vision, determining the shape, location or quantity of appearance is the one of major focus and concern in many applications. In our project, for human detection using a circular marker various numerous techniques were openly viable, but we opted the most robust and commonly used technique of Circular Hough Transform (CHT).

The Hough Transform is used for determination of the parameters of a circle given that a number of known points fall on the parameter. The parameter space is 3D, if the circles in an image are of known radius R, then the search can be reduced to 2D. The circle with Radius R and a center (a, b) can be expressed using the parametric equations as follows:

$$x = a + R cos\theta$$

$y = b + R sin\theta$ The locus of (a, b) points in the parameter space fall on a circle of radius R centered at (x, y). The true center point will be common to all parameter circles, and can be found with a Hough accumulation array. While the angle arches through the full 360 degree range the coordinates (a, b) trace the circumference of the circle. A search program is then used to find the triple parameters to give the description of the circle i.e. R, a, b in the image containing many points, some of which fall on the perimeter.

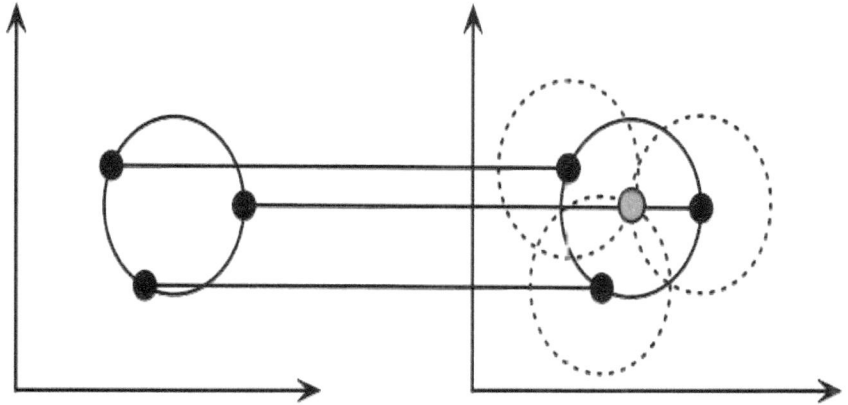

Figure 15 each point in geometric space (left) generates a circle in parameter space (right). The circles in parameter space intersect at the center (a, b)

A three dimensional array is used by the circle Hough transform technique where the first two elements represent the coordinates of the center of circle while the third is the value of radius. The values in the array keep increasing every time a circle is drawn. The accumulator keeps counting the number of circles passing through the coordinates of each edge point that proceeds to a vote to find the highest count and the circles seen in images are those coordinates which have highest count.

In our project, R is not predefined or fixed; it varies with forward and backward motion of caretaker. Since caretaker is assumed to be within 2-3m range of the wheelchair, so we measured radius for nearest and farthest point. The fact that the parameter space is 3D makes a direct implementation of the Hough technique more expensive in computer memory and time. Limiting range of radius increased speed.

Multiple circles with the same radius can be found with the same technique. The center points are represented as red cells in the parameter space drawing. Overlap of circles can cause spurious centers to also be found, such as at the blue cell. Spurious circles can be removed by matching to circles in the original image.

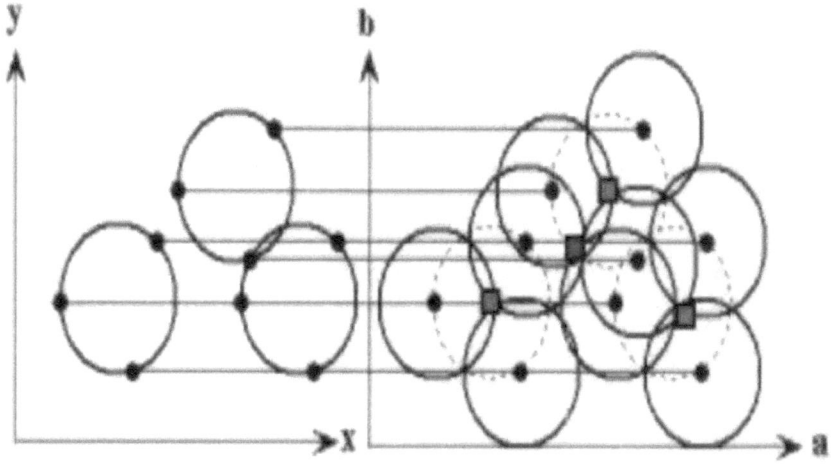

Figure 16 each point in geometric space (left) generates a circle in parameter space (right). The circles in parameter space intersect at the center (a, b)

Since the radius is not fixed, then the locus of points in parameter space will fall on the surface of a cone. Each point (x, y) on the perimeter of a circle will produce a cone surface in parameter space. The triplet (a, b, R) will correspond to the accumulation cell where the largest number of cone surfaces intersect. The drawing below illustrates the generation of a conical surface in parameter space for one (x, y) point. A circle with a different radius will be constructed at each level, r. The search for circles with unknown radius can be conducted by using a three dimensional accumulation matrix.

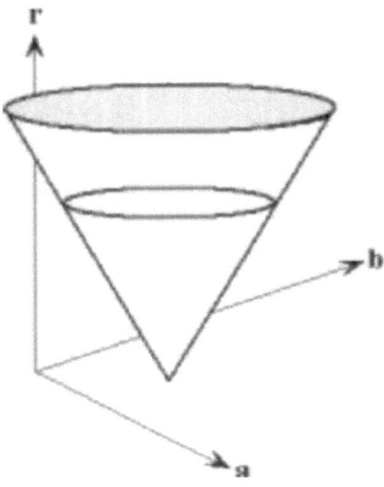

Figure 17 cone formed b considering all possible values of radii.

Here is the end result, obtained by this transform:

Figure 18 Circles detected using Hough circle transform.

3.3.5.1 Amendments in the algorithm

Usually circle Hough transform technique is applied on pictures, but we have used for live video coming from the camera. Applying whole algorithm on each and every frame of the video was time consuming. Also the variation in the results was occurring so fast that it became impossible for wheelchair to properly track the target. So we applied Hough transform on multiple frames, took the mean of the radius and center of the recognized circle and used these mean values to command wheelchair.

3.3.6 Obstacle Detection

Laser range finder (LRF) is used to detect obstacles. A laser rangefinder is a rangefinder which uses a laser beam to determine the distance to an object. The most common form of laser rangefinder operates on the time of flight principle by sending a laser pulse in a narrow beam towards the object and measuring the time taken by the pulse to be reflected off the target and returned to the sender. By using the relation

$$distance = velocity * time$$

it computes the distance from objects around LRF. Objects present in front of LRF, with distance less than a threshold value, are treated as obstacles.

3.3.7 Obstacle Avoidance

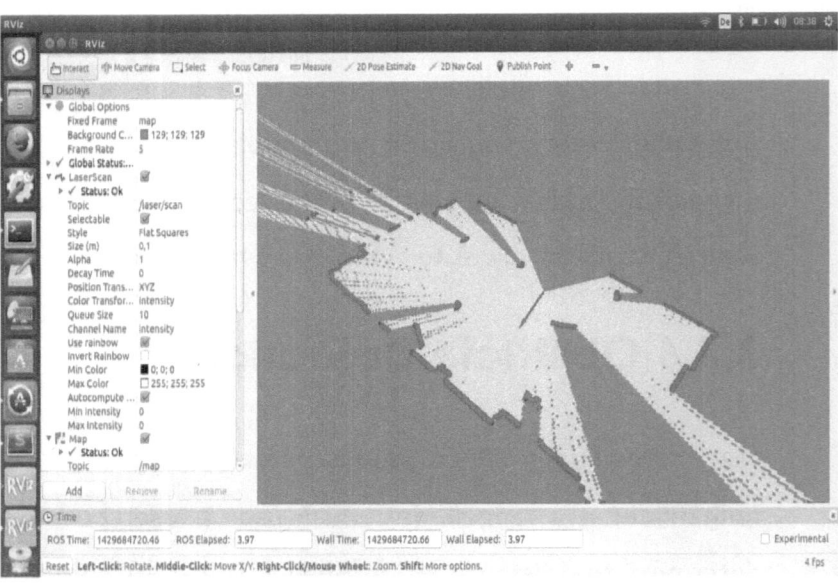

Figure 19 View of data gathered by LRF

Basic Obstacle Avoidance

In this approach, the wheelchair stops whenever an obstacle is detected with distance less than a threshold value.

Advanced Obstacle Avoidance

In advanced obstacle avoidance technique, wheelchair moves along the boundary of the obstacle, and once obstacle is avoided, it rotates at 360° and finds the caretaker; once found it starts tracking it again.

3.3.8 Velocity Calculations

With the processed data of laser range finder and camera, now we are in position to make decisions for the movement of wheelchair and we can command it accordingly. Multiple situations that might occur along with decisions in those cases are listed here:

3.3.8.1 Angular motion

- target person towards the left from the center

    When seen through camera if the target person is present towards the left from center of the image (with distance between center and person more than a threshold value), we should move wheelchair towards left, till the time when target is in the center of the visual range.

- target person towards the right from center

    When seen through camera if the target person is present towards the right from center of the image (with distance between center and person more than a threshold value), we should move wheelchair towards right, till the time when target is in the center of the visual range.

3.3.8.2 Linear motion

Now since the target is in center i.e. wheelchair is heading rights towards the target (caretaker), we have to decide the linear motion of the wheelchair. It is also based on the camera feed. Wheelchair is aware of the size of the radius of the circle (tag present on shirt of caretaker) when wheelchair is within 1m meter range from caretaker, so it will remain stationary. As the caretaker will start moving, the radius of the circle will start decreasing/increasing based on forward/backward motion. If caretaker is moving ahead wheelchair will follow its caretaker, while in case if caretaker is moving back, wheelchair will remain stationary.

3.3.8.3 Obstacle avoidance

While wheelchair is following the caretaker many obstacles will come in the way, whenever there is such case, wheelchair will start following the path planned and after avoiding the obstacle it will continue to follow the caretaker again.

3.3.9 Software Architecture

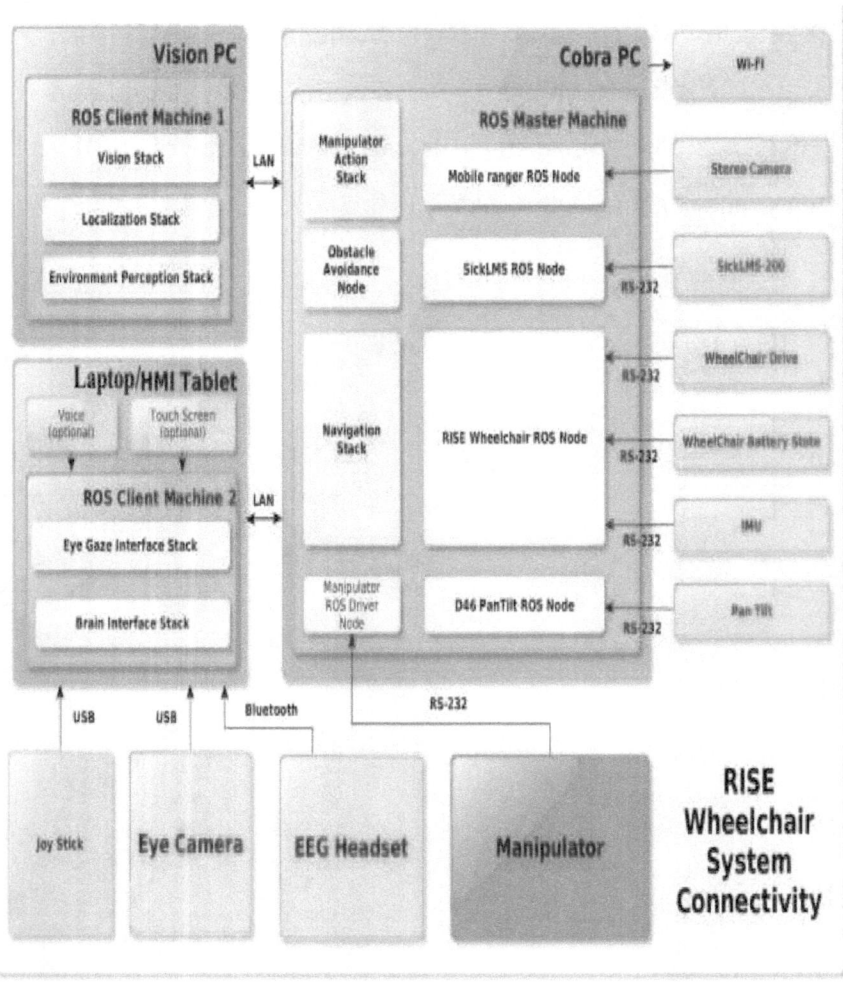

Figure 20 Software architecture

3.4 Operation

By now we have explained everything about wheelchair, from scratch; presenting things step by step to make a strong foundation of concept in the mind of reader. Now let's see how this Wheelchair can be controlled. Given below is complete description of operation of wheelchair, follow it step by step to avoid inconvenience.

3.4.1 System Requirements

In order to connect to wheelchair and control it, here are some requirements that your system must fulfill:

- Ubuntu 14.04 LTS
- ROS
- OpenCV
- Wi-Fi

Note: To execute the following commands you need to configure your own network settings. Then enter the wheelchair and edit the /etc./hosts file and add your own network configuration settings (IP address + PC name) in the file. Then restart your network to apply changes.

1. Connect to Wi-Fi.
2. Start terminal and write
3. $ ssh 10.12.42.190 –l mobirobo –X (Connects to the wheelchair)
4. $ cd ROS
5. $ sudo./startup.sh (This file configures the serial port connections with the motors.)
6. $ roslaunchrisewheelchairewchair_protected.launch (turns on LRF and ROS master)
   If it doesn't work execute $ export LC_ALL="en_US.UTF-8"
   then repeat the previous step.
7. New terminal
8. $ ROS_MASTER_URI=http://10.12.42.190:11311 (Sets ROS master on remote PC)
9. $ rosservice call /Robot/rise.../motors_enable

10. $ rostopic pub /cmd_velgeometry_msgs/Twist –r 40 – '[0.7, 0.0, 0.0]' '[0.0, 0.0, 0.0]' (moves the motor forward)

Page intentionally left blank

Chapter No. 4

EXPERIMENTS AND RESULTS

4.1 Experimentation

We have tested this project for multiple scenarios based on type on motion of caretaker, timings of the day, indoor/outdoor etc.

4.1.1 Type of motion of caretaker

1. Stationary

    Whenever the wheelchair detects the marker from its camera it generates a forward command based on the radius of the marker. Currently the condition on the radius is 100px. If the radius is less than 100px then the wheelchair moves towards the marker.

    Linear forward motion

2. Angular motion

    The code uses an alignment function which aligns the webcam with the marker center. As the marker moves towards a side the wheelchair turns in the corresponding direction to align with the marker.

3. Both linear and angular motions, alternatively.

In this scenario, when caretaker is moving both in forward and sidewise direction, wheelchair generates angular and linear velocity commands simultaneously.

4.1.2 Indoor/outdoor

1. Indoor
   Wheelchair perfectly tracks caretaker in indoor environment.
2. Outdoor
   In outdoor environment, performance depends on light intensity incident on the marker.

4.1.3 Tele-operation using Android App

There are no limitations in this approach; wheelchair accurately follows velocity commands generated by teleop app.

4.1.4 Keyboard Teleop

Again same results, wheelchair corresponds immediately to keyboard key click, and moves in corresponding direction.

4.2 Limitations

- It works only if the illumination on the tag is uniform; in case of high intensity light, unexpected results may occur. That's why it is recommended for indoor use only.
- Techniques used assume that the caretaker is within 5m range from wheelchair, which is a sufficient condition for this particular application.

Chapter No. 5

CONCLUSION

In this book, we have introduced a new way of human detection and following. It detects a person based on the tag printed on its shirt. Based on the features collected via camera, it measures the distance and location of target human; and generates corresponding velocity commands to the motors of the wheelchair. While following a person there might come an obstacle in the way, using LRF it detects the obstacle, and moves along a new planned path, so that wheelchair can follow its caretaker without hitting any hurdle in its way. These algorithms were combined and implemented on wheelchair of RISE Lab, SMME; results proved that it can be used on commercial scale.

5.1 Future Work

- Our focus was on static obstacles; the very same approach can be extended and applied on moving obstacles as well.
- The tag on the shirt of caretaker can be made to include some text i.e. serial number of the chair, so

that wheelchair crowded environment every chair can recognize its caretaker.

- The task done now, uses webcam of laptop, instead of this, already mounted camera, or a new camera can be added to it, for permanent and dedicated use.

REFERENCES

[1] SongminJia, Liang Zhao, Xiuzhi Li, Wei Cui, Jinbo Sheng, "Autonomous Robot Human Detecting and Tracking Based on Stereo Vision", Proceedings of the 2011 IEEE International Conference on Mechatronics and Automation, August 7 – 10, 2011.

[2] SongminJia, Liang Zhao, Xiuzhi Li, "Robustness Improvement of Human Detecting and Tracking for Mobile Robot", Proceedings of 2012 IEEE International Conference on Mechatronics and Automation, August 5 – 8, 2012.

[3] MehrezKristou, Akihisa Ohya and Shin'ichiYuta, "Panoramic Vision and LRF Sensor Fusion Based Human Identification and Tracking for Autonomous Luggage Cart", 18th IEEE International Symposium on Robot and Human Interactive Communication, Sept. 27-Oct. 2, 2009.

[4] MehrezKristou, Akihisa Ohya and Shin' chiYuta, "Target person identification and following based on

omnidirectional camera and LRF data fusion", 20th IEEE International Symposium on Robot and Human Interactive Communication, July 31 - August 3, 2011.

[5] Nicola Bellotto and Huosheng Hu, "People Tracking and Identification with a Mobile Robot", Proceedings of the 2007 IEEE International Conference on Mechatronics and Automation, August 5 - 8, 2007.

[6] OrasaPatsadu, ChakaridaNukoolkit and BunthitWatanapa, "Human Gesture Recognition Using Kinect Camera", 9th International Joint Conference on Computer Science and Software Engineering, 2012.

[7] Jia, S., Zhao, L., & Li, X. 2012. Robustness improvement of human detecting and tracking for mobile robot. In Mechatronics and Automation (ICMA), 2012 International Conference .pp. 1904-1909. IEEE.

[8] Kristou, M., Ohya, A. &Yuta, S. 2011. Target person identification and following based on omnidirectional camera and LRF data fusion, 20th IEEE International Symposium on Robot and Human Interactive Communication. Pp.419-424.

[9] Bellotto, N. & Hu, H. 2006. Vision and laser data fusion for tracking people with a mobile robot,

Proceedings of the 2006 IEEE International Conference on Robotics and Biomimetics. Pp.7-12.

[10] Patsadu, O., Nukoolkit, C. &Watanapa, B. 2012. Human gesture recognition using kinect camera, International Joint Conference on Computer Science and Software Engineering. pp. 28-32.

[11] Petrović, E., Leu, A., Ristić-Durrant, D. &Nikolić, V. 2013. Stereo vision-based human tracking for robotic follower International Journal of Advanced Robotic Systems. 10:230 DOI: 10.5772/56124

[12] D. Schulz, W. Burgard, D. Fox, and A. B. Cremers, "People tracking with mobile robots using sample-based joint probabilistic data association filters,"Int. Journal of Robotics Research, 2003.

[13] E. A. Topp and H. I. Christensen, "Tracking for following and passing persons" inInt. Conference on Intelligent Robots and Systems (IROS), 2005.

[14] R. Gockley, J. Forlizzi, and R. Simmons, "Natural person-following behavior for social robots," inInt. Conference on Human-robot Interaction, 2007.

[15] S. Hemachandra, T. Kollar, N. Roy, and S. Teller, "Following and interpreting narrated guided tours," in

Int. Conference on Robotics and Automation (ICRA), 2011.

[16] K. O. Arras, B. Lau, S. Grzonka, M. Luber, O. M. Mozos, D. MeyerDelius, and W. Burgard, "Range-based people detection and tracking for socially enabled service robots," in Towards Service Robots for Everyday Environments, pp. 235–280, 2012.

[17] M. Kobilarov, G. Sukhatme, J. Hyams, and P. Batavia, "People tracking and following with mobile robot using an omnidirectional camera and a laser," in Int. Conference on Robotics and Automation (ICRA), 2006.

[18] SambartaDasgupta&Swagatam Das, 'Automatic circle detection on digital images with an adaptive bacterial foraging algorithm'/ October 15, 2009

[19] T. D'Orazio,C. Guaragnella,M. Leo,A. Distante,' A new algorithm for ball recognition using circle Hough transform and neural classifier,Volume 37,Issue 3,March 2004

[20] Victor Ayala-Ramirez & Carlos A Gracia,'Circle detection on images using genetic algorithms, University of Gwantawa ,Electronics and

communication engineering department, Mexico/November 28,2005

[21] T J Atherton & D J Kerbyson. 'Size Invariant Circle Detection', Department Of Computer Science, University Of Warwick, Coventry CV4 7Al UK, /February 10, 1997

[22] D. Montemerlo, S. Thrun, and W. Whittaker, "Conditional particle filters for simultaneous mobile robot localization and people-tracking," in Int. Conference on Robotics and Automation (ICRA), 2002.

[23] M. Wang and J. N. K. Liu, Fuzzy logic-based real-time robot navigation in unknown environment with dead ends, Robot. Autonomous Syst., vol. 56, no. 7, pp. 625643, 2008.

[24] M. Lawn, & T. Takeda, 'Design of a robotic-hybrid wheelchair for operation in barrier present environments', Proceedings of the 20th Annual International Conference of the IEEE Engineering in Medicine and Biology Society, Vol. 20, No 5

[25] T. Röfer, & A. Lankenau. 'Ensuring Safe Obstacle Avoidance in a Shared-Control System', Proceedings of the IEEE/RSJ/GI International Conference on

Emerging Technologies and Factory Automation 1999, Vol. 2, 1405-1414, October 1999.

[26]   S. Fioretti, T. Leo, & S. Longhi, 'A Navigation System for Increasing the Autonomy and the Security of Powered Wheelchairs', IEEE Transactions on Rehabilitation Engineering, Vol. 8, No. 4, 490- 498, Dec 2000.

[27]   E. Prassler, J. Scholz, M. Strobel, & P. Fiorini, 'An Intelligent (Semi-)Autonomous Passenger Transportation System', Proceedings of the IEEE/IEEJ/JSAI International Conference on Intelligent Transportation Systems Proceedings 1999, 374-379, October 1999.

[28]   J. Hockenberry, (cited 20-May-2001) 'A revolutionary new wheelchair', NBC http://www.msnbc.com/news/285231.asp, June 2000.

[29]   G. Pires, R. Araujo, U. Nunes, & A. Almeida, 'RobChair-a powered wheelchair using a behavior based navigation', International Workshop on Advanced Motion Control 1998, 536-541, June 1998.

The End

www.ingramcontent.com/pod-product-compliance
Lightning Source LLC
Chambersburg PA
CBHW020445220526
45464CB00002B/870